Copyright © 2024 by - Kristen Lenders
All Rights Reserved.
It is not legal to reproduce, duplicate, or transmit any part of this document in either electronic means or printed format. Recording of this publication is strictly prohibited.

Dedication

Written for all those who love Riese. Written to remember the love he shared with us, for those who never want to forget the special man he became, but especially for Kolten Riese and the nieces and nephews that Riese never had the opportunity to enjoy time with on earth.

About the Author

Kristen is a native of the Palos Verdes Peninsula in Southern California. She loves spending time outdoors and has shared her love for it with her family. She met and married her high school sweetheart in the 80s and they have three redheads. They have recently entered the joyous new world of being Grandparents!

Uncle Riesey-Boy
Love like Riese

By:
Kristen Lenders

When Riese was a baby, his daddy would make up bedtime stories for him, his older brother and older sister. Riesey-Boy was always the hero.
When he was your age he especially loved to dress up and say, "I'm Moo Man!"
(He couldn't say Superman yet. He was still too little.)

Riesey-Boy loves to smile and have fun but he especially loves to make others smile and have fun. Riesey-Boy soon became everyone's favorite Uncle!

Uncle Riesey-Boy loves exploring and playing outdoors in God's beautiful creation. But he especially loves exploring and playing outdoors with the people he loves! Do you?

Uncle Riesey-Boy loves making friends with furry animals and creepy crawly ones, too. But he especially loves playing with them! Do you, too?

Uncle Riesey-Boy loves outdoor adventures like hiking the tallest mountains and surfing the deepest oceans. But he especially loves doing these things with new friends. Will you, too?

He loves boating in lakes and skiing through big trees. But especially loves doing these things on a family vacation! Do you?

Uncle Riesey-Boy loves to drive cars and sail all kinds of boats. But he especially loves flying airplanes! Maybe you will, too!

He loves dancing at weddings and hoedowns. But he especially loves showing people how he loves doing "the Worm!"

Because Riesey-Boy loves God so much, he especially loved it when God asked him to keep flying to heaven! Just imagine the great adventures he is having right now with his big family and new friends up there!

Psalm 36:5
Your steadfast love, O Lord, extends to the heavens, your faithfulness to the clouds.

In loving memory of the six on Cessna 550 Citation II, July 8, 2023.

"There are many reasons to be sad, but there are more reasons to be happy!" ~Uncle Riesey-Boy

IYKYK

Printed in the USA
CPSIA information can be obtained
at www.ICGtesting.com
LVRC091300150924
790971LV00011B/373